INERTIA

Rebecca Woodbury, Ph.D., M.Ed.

Gravitas Publications Inc.

INERTIA

Illustrations: Janet Moneymaker

Inertia
ISBN 978-1-950415-22-9

Published by Gravitas Publications Inc.
Imprint: Real Science-4-Kids
www.gravitaspublications.com
www.realscience4kids.com

RS4K

Photo credits: Cover, Title Pg, P.7: By mslok, AdobeStock; P. 5. By Africa Studio, AdobeStock; P.9. By JenkoAtaman, AdobeStock

How do things move?

Use arms and legs?

How does a

soccer ball move?

It gets kicked?

How does a marble move?

I don't know.
Do you?

How does a bicycle move?

How does it move?

Aristotle was a Greek philosopher. He thought **force** made things move.

FORCES KEEP THINGS MOVING!

Review: FORCE

In physics...

Force is any action that changes:

- The **location** of an object.

- The **shape** of an object.

- **How fast or how slowly** an object is moving. (This is called the **speed** of an object.)

But the Italian scientist
Galileo found that
Aristotle was wrong!

Aristotle was wrong!

- 13 -

Inertia is when an object resists a change in motion.

Do I have inertia when I sit?

Maybe.

For example...

Have you ever wanted to keep playing instead of going inside to eat dinner?

I have!

That's inertia!

Because you did not want
to change your motion
(playing) to a new motion
(going to dinner) ·······················➤

That's inertia!

Inertia keeps things that are in motion moving.

Inertia also keeps objects that are not moving at rest.

Force changes the direction of an object?

Yes! And force is what stops an object.

SO IT IS INERTIA THAT KEEPS OBJECTS MOVING, NOT FORCES.
YES! THAT IS RIGHT!

How to say science words

Aristotle   (ER-uh-stah-tuhl)

force   (FAWRS)

Galileo   (gaa-luh-LAY-oh)

inertia   (ih-NER-shuh)

location   (loh-KAY-shun)

physics   (FIH-ziks)

shape   (SHAYP)

speed   (SPEED)